Android Programming For Beginners:

The Ultimate Android App Developer's Guide

By

Joseph Joyner

Table of Contents

Android Programming For Beginners: The Ultimate
Android App Developer's Guide

By Joseph Joyner

First Published, 2015

Printed in the United States of America

Introduction

Mobile application development is now the hottest thing in the programming world. One is the Android and another is iOS. Although the numbers tend to vary frequently, it appears that Android has grabbed the topmost place from Apple.

Chapter 1. Android Basics

Android is a Linux based mobile operating system that was originally developed by same name Android, Inc. In 2005, Google purchased Android and continued its development work through its efficient worker.

Each android application has four unique components. By combining these 4 components, it is possible to create android application.

Activities

An Android operating system, an activity is any lone screen with a UI (User Interface). An excellent example is an email application where one activity shows you the inbox, another activity is used to create new emails, and other activity permits you to read emails. For the inexperienced, this may appear like extra work. This makes your job truly as an Android programmer or developer much easier because the modular drawing and construction helps you to change one feature of the application without disturbing or affecting the rest. For example, you may desire to have a different color scheme when users generate a new email that actually differs from colors that is used in the rest of the app. By using activities this does a very simple task.

Services

This component provides one of the most important features in android. Services are run in the background. They are liable and responsible for long-running operations or performing some tasks that does not need user interaction. Services are not allowed to have a user interface. When you use android and music player on your phone, this often runs as services in the background. So, the music will continue running even when you switch to a different part of the mobile.

Another admired service implementation is getting background data. For example, if you use Gmail, you are acquainted with push notifications. A service that runs in the background verifies for new mail once in a while without manual involvement from the user. This is an ideal example of a service.

Content Providers

Content providers help handle application data of its own. In Android, data can be easily stored within the file system, in a SQLite database, on different Web, or in other local storage locations for example; a MicroSD card. Contact-that exposes user numbers and other information to other applications is the content provider example. Contact information that is stored in the "Contacts" app can be easily queried. This application is capable of reading or writing information to the database using content providers.

Broadcast Receivers

Broadcast receivers are responsible for acting in response to system broadcast announcements. Although most broadcasts are started with the Android operating system, your application can create broadcasts as well that allow other system applications to recognize what is going on. It is also responsible for "listening" for other broadcasts and taking some action when a broadcast is good or detected. This could be as easy as status bar notification or it could initiate another service or activity automatically.

Android Platform

Android uses Java for application development. Java is an object oriented language where object is the life of programming. To code in android, one must know some basics of Java. Without Java, you can't perform or do logical operations. Using Java API, you can code your Android apps that are provided by Google, and it actually compiles into class files. The resemblance ends here, Java Virtual machine (JVM) is not used by Android for executing class files, in its place, it uses a Dalvik virtual machine. This is not a true JVM and doesn't function on Java byte code. To run on (DVM), class files are again compiled in DEX format. After conversion to this format, class files together with other resources are combined in Android Package (APK) for installation and distribution into various android devices. But it is important to know that Dalvik VM doesn't support and maintain all J2SE API. If you are determined to use Eclipse IDE for coding Android Applications, then you don't have to worry much because, it will assist you with code finishing point. Now let's see how android runs on the device.

Chapter 2. How Android App Runs on Android Device?

If you are acquainted with Linux and idea of process, then it's very easy to understand how android app runs. By default, each application is assigned a single user id that is unique by the Android operating system. After initiating android application, they run on their self process, within their own virtual machine. Android manages initiating and shutting down the whole application process, whenever needed. This means, each and every android application run in separation with other, but they can surely request entrée to hardware and other system resources. Without these, it's impossible to run in android smart phone. If you know something about J2ME, then you may discern about permissions to complete some tasks. So when an android application is started or installed, it requests necessary permission needed to connect phone book, internet and other system resource. User openly provides grant these permissions, consents or it may reject. All these authorizations or permissions are declared in manifest file. Unlike Java, Android manifest is one type of XML file, which actually lists all the parts of apps, and settings for those mechanisms.

Android Architectures

The Android operating system is one type of a load of software components which is generally divided into 5 sections and 4 main layers. They are described in the following:

Linux kernel

At the base of the layers is Linux - Linux 2.6 with just about 115 patches. This offers basic system functionality like memory management, process management, device management like keypad, camera, display, etc. Also, the kernel deals with all the things that Linux is truly good at, such as a vast array of device drivers and networking, which seize the pain out of interfacing to marginal hardware.

Libraries

At peak of Linux kernel there is a place of libraries containing open-source Web browser engine, well known library libc, an efficient database SQLite database which is a helpful repository for both sharing and storage of application data, libraries to have fun, record audio and video, SSL libraries are liable and responsible for various Internet security, safety, preservation etc.

Android Runtime

This is actually the 3rd section of the architecture of the android and accessible on the 2nd layer from the base. This section offers a key part called Dalvik Virtual Machine (DVM) which is a type of JVM specially planned, designed, considered as well as optimized for Android.

The Dalvik Virtual Machine takes the advantages of Linux core features like multi threading and memory management, which is essential in the Java language. It enables every application to run in its own method, with its own occurrence of the Dalvik virtual machine (DVM).

The Android runtime gives a set of interior libraries which allow Android application developers to inscribe Android mobile applications using typical Java programming language.

Application Framework

This layer provides many higher-level protections and services for applications in the shape of Java classes. Application developers are permitted to make use of these facilities in their applications.

Applications

You will get the entire Android app at the top layer. You will inscribe your application to be set up on this layer only. Examples of these applications are Browser, Contacts Books, Games etc.

Chapter 3. Android Development Tools

Android SDK

Android SDK consists of all the necessary tools to make, compile and package Android mobile applications. Most of these instruments are command line based. The main way to develop Android mobile applications is based on the most popular object oriented language Java.

Android Debug Bridge

It's one of the most interesting part to know. The Android SDK that includes the Android debug bridge which is important to test and run a virtual device is simply a tool that lets you to join to a virtual or genuine Android device, for the reason of managing the device or debugging or correcting your application.

Integrated Development Tools

Google gives two efficient integrated development environments (IDEs) to enlarge and develop new applications. To create applications and create new developers with immense opportunities, Google develops an IDE. It is called Android Studio. This IDE is actually based on the IntelliJ IDE. ADT or Android Development Tools are rooted in the Eclipse IDE. It has some components, which expand the Eclipse with huge opportunities and Android development abilities.

Both IDEs have all needed functionality to make, debug, compile and deploy Android Mobile applications. They also permit the developer to produce and start virtual Android devices for testing. So, you don't need to pass your app every time for the testing. You have your virtual device and do everything you want. Currently the ADT tooling frequently uses a particular Eclipse build system without taking the new Gradle build system. This can make discrepancies in your build. If you feel like to have the Gradle support for Eclipse, please contact and go to Support Gradle builds and press the star or initiate sign.

Since Android specific files, both tools offer specialized editors. The majority of Android's configuration files are written in XML. In this reason, these editors let you to switch between the XML depiction or representation and a structured UI for entering the data. This explanation tries to explain the usage of

Android Studio and the Eclipse based Android Development Tools (ADT) tooling.

Android RunTime (ART)

Android Run Time is an important feature in android. Android 5.0 first uses the Android RunTime (ART) for all Android applications. Ahead Of Time compilation is the process of running ART. During the deployment procedure of a mobile application on a smart phone Android device, the code is converted into machine code which results in approximately. 30% larger compile the code, but permits faster execution from the start of the application. This saves battery life as well, as the compilation is merely done once, during the initial start of the android application.

The dex2oat tool deal with the .dex file that is created by the Android tool modify and accumulates that into an efficient format called Executable and Linkable Format (ELF file). It contains the dex code, compiled local code and meta-data. Maintaining the .dex code lets that existing tools still do some work. The garbage compilation at an Android Run Time has been optimized to decrease times in which the application freezes.

Developing Android applications

Applications are mainly written in the most secured object oriented language called Java. During progress, the developer makes the Android detailed configuration files and handles the logical operation in the Java programming language.

The Android Studio tools or the ADT alter these application files, visibly to the user, into an Android mobile application. When developers generate the use in their IDE, the entire Android application is compiled, fully packaged, deployed and started.

Conversion process

The Java source files are directly converted to Java class files. This is easily done by the Java compiler. The Android SDK consists of a tool called dx which translates and converts Java class files into a .dex file. The entire class files of the application are found in this .dex file. During this exchange process superfluous information is optimized in the.dex file. For instance, if the same String is found in dissimilar class files, the .dex file holds only one mention of this String.

These files are consequently much smaller in amount than the equivalent class files. These files and the entire resources of an Android project, for example, the XML files and images, are packed and combined into an .apk (Android Package) file. This operation is totally performed by aapt. This resulting .apk can be used to run in various android mobile devices because this .apk file combines all the necessary data to run on every Android device including its version.

Chapter 4. Security and Permissions

Security concept in Android

You are very happy to know that the Android system installs each Android application with a unique user plus group ID. Every application file is personal to this generated user, e.g., another application can't access these files. Besides, each Android application is started in its own procedure.

For that reason, by means of the underlying Linux kernel, all Android applications are isolated from supplementary running applications. If data would be shared, the application must do this openly via an Android component which can handle the sharing of the data, such as via a service.

Permission concept in Android

Android includes a permission system with predefines permissions for certain tasks. Each application is able to request required permissions and also define some new permission. For instance, an application may say publicly that it needs access to the Internet.

Permissions have some different levels. Some permission is automatically rejected by the Android system, several are automatically granted. In the majority cases the requested permissions are obtainable to the user prior to installing the application. The user wants to decide if these permissions shall be known to the application.

But the user wants to deny a required permission; the related application can't be installed. Then the check of the permission is just performed for the period of installation; permissions can't be denied or granted after complete the installation. An Android application also asserts the required permissions in its AndroidManifest.xml layout configuration file. It can also define extra permissions which it can use to limit access to certain components.

Chapter 5. Needed Tools for Environment Setup

You are very happy to know that android application development can be started through different operating systems (OS). It will be possible to code in Microsoft Windows XP or the latest version of Microsoft Windows. Besides, it can be also started on Linux, including GNU C Library 2.7 or later and started with Mac OS X 10.5.8 or any later version with Intel chip.

One more reason is that to develop and create android application you will need some development tools or software like Eclipse, Android Studio etc. You are very happy to know that you can get them or download them freely. Google has made it simple to use for everyone. This book will show you how to develop android application using Eclipse IDE. Other supported tools for Eclipse are given in the following:

A. Java JDK5 or JDK6

B. Android SDK

C. Android Development Tools (ADT) Eclipse Plugin (optional)

Setup Java Development Kit

If you want to download the latest version of the Java Development Kit, you can download it from Oracle's Java site, namely Java SE Downloads. Besides, You will get more instructions for installing the SJDK in downloaded files and also maintain the certain instructions to install as well as configure the setup. As a final point set PATH as well as JAVA_HOME environment variables to submit to the directory that includes Java. If you are using Windows operating system with installed the Java Development Kit in C:\jdk1.6.0_15, you would have to take the subsequent line in C:\autoexec.bat file.

Otherwise, you must follow the direction. At first right-click on My Computer, next select Properties, after that Advanced, then Environment Variables. At that moment, you would update the PATH value and must push the OK button.
Otherwise, if you can use an IDE Integrated Development Environment Eclipse, then it will

automatically identify where you have installed your Java.

Set Up Your Android SDK

It is possible to download the newest version of Android SDK from different websites including Android official website. If you are installing or setting up SDK on Windows computer, after that you will get an installer called _rXX-windows.exe, so then just try to download in a hurry and apply this exe will start Setup wizard to show you all through of the setting up, so just pursue the instructions cautiously. Lastly, you will get Android SDK Tools set up on your computer. If you are setting up SDK on Linux or Mac OS, confirm the instructions given together with the rXX-macosx.zip for android_sdk for Mac and _rXX-linux.tgz for android-sdk for Linux. This lesson will believe that you will setup your environment on a Windows machine if you have Windows 7 operating system. So let's open Android SDK Manager uses the selection All Programs then Android SDK Tools and then SDK Manager.

Once you have opened SDK manager, it's time to set up other needed packages. It will list down entirety 7 packages to be installed by default, but I will propose to de-select Samples for SDK packages and documentation for Android SDK to decrease installation time. After that, click Install 7 Packages button to continue. If you agree to set up all the packages, choose Accept All radio button and continue by clicking Install button. At this moment, let SDK manager perform its work and you leave, wait until all the packages are entirely installed. It may take a while depending on your connection of internet. Once all the packages are installed completely, you can exit from SDK manager using the top-right cross button.

Set Up Your Eclipse IDE

All the instances in this book have been written and shown using Eclipse IDE. So we would propose you should have the most recent version of Eclipse set up on your machine. For Eclipse IDE to be installed, download the latest Eclipse from different websites. Once you have completed the downloading process, unload the binary distribution into a suitable location. For instance in /usr/local/eclipse on Linux, or C:\eclipse on windows and lastly set PATH variable properly. Eclipse can be initiated by executing the subsequent commands on a windows machine, or you can merely double click on eclipse.exe. The commands are:

%C:\eclipse\eclipse.exe

Again, eclipse is able to be initiated by performing the commands on Linux machine that are given in the following:

$/usr/local/eclipse/eclipse

After a flourishing startup, if everything is well then it should establish the Eclipse IDE without any error viewing in the display.

Prepare your Android Development Tool Plug-in

This will help you setting up your ADT plug-in for Eclipse. If everything goes well, you can start by launching Eclipse IDE and choose Help > Software Updates > Install New Software

After entering a location and ADT plug-in as name by using add button, you can then click ok. The Eclipse will start searching and list down all the plug-ins. After choosing all of them, you can install easily.

Create Android Virtual Device

To judge or test your android application you are going to need a virtual device which will show you the results of your performance. To create, you have to launch AVD using Eclipse menu Window>AVD Manager >. Then Use New button to generate a fresh or new Virtual Device and enter the needed information, before clicking Create AVD button.

If everything goes well, your AVD will be created successfully; it means you have completed everything to create your first android application. After completing all your steps for setting up your environment, it's recommended to close Eclipse IDE and everything and restart your machine. Now, you are ready to start building your first android app.

Chapter 6. Creating Android Application

The 1st step is to make a simple Android Application. To create this app, we will show you using Eclipse IDE. Then follow the option File -> New -> Project and lastly select Android New Application wizard. Now write the name of your application as HelloWorld.

Before you start to run your first app, you should be careful of the following directories, files and folders.

1. src: This is where java code actually resides and writes. This holds the .java source files for your entire project. By default, it has a MainActivity.java source file holding an activity class that actually runs when your app is started by clicking the app icon.

2. gen: This is actually the most important file in building android application. But it shouldn't be modified. When you enter a button or text view or anything, a number is created to identify those things uniquely. This actually holds the .R file which is a compiler-generated file that mentions all the resources and keys found in your project. You should not and must not change this file.

3. bin: this is the folder where .apk files can be found which is built by the ADT during the build time and everything else required to run an application.

4. res/drawable-hdpi: It is one type of directory for drawable objects that are specially designed for high-density screens or displays.

5. res/layout: It is also a directory for files that describe your app's UI.

6. res/values: This is another directory for other different XML files that have a collection of resources, such as colors and strings definitions.

7. AndroidManifest.xml: AndroidManifest.xml is the manifest or apparent file which explains the fundamental characteristics of the application and describes each of its parts.

The Main Activity File

MainActivity.java is a Java file which is the life of android mobile application. This is the real application file which eventually gets translated and converted to a Dalvik executable and also runs your app. Here you will see some default code that is automatically generated by the application wizard: First you have to include the package name.

package com.example.helloworld;

Package declarations and other imported classes will be needed to work efficiently and properly. The package name must be the same when you created the new android project. Imported classes are necessary because you will use some references or classes that will be needed to be imported. Some examples are like this:

import android.os.Bundle;

After that, the code public class Main Activity extends Activity means you are inheriting Activity class which is public from android.app.activity class.

Then you have to define bundle oncreate method because when the activity is started then oncreate method is invoked.

You can write code as you wish in the main method to experiment. But you have to remember something when coding. You must have some concept of java because it's quite impossible to be an android programmer without the knowledge of java.

The Manifest File

When you are going to develop an android application you need to declare all of your components in a file to introduce it to the android operating system. This file is called manifest file and this is denoted as AndroidManifest.xml which resides in the source of the application project index. This file performs as an interface between the Android operating system and your application, so if you don't declare your part in this file, at that moment it will not be recognized and considered by the operating system. In a default manifest file, you will see package name, android sdk minimum version and target version. This will define in which android version will it support.

Here <application>...</application> tags are the most important and significant. Because within these tags all the components related to the components lie here. The feature android:icon will indicate to the icon that is existing under res and then drawable-hdpi. Again the application takes the image called ic_launcher.png which canbe found in the drawable folders. To declare activity you can declare within the activity tag and android:name attribute indicates the fully competent class name of the Activity subclass. Again the android:label feature specifies a string to identify as the label for the activity. If you have multiple activities, you can declare them within the activity tags.

The main action for this intent filter is called android.intent.action.MAIN to specify that this activity provides as the entry point for the android application. The type for the intent-filter is called android.intent.category.LAUNCHER to show that the application can be started from the device's launcher icon.

Again, the @string signifies to the strings.xml file described below. Hence, @string/app_name indicates to the app_name string declared in the strings.xml file "HelloWorld". Other strings get settled in the application.

The Strings File

This strings.xml file is found in the res/values folder and it has all the text that your application utilizes and uses. For instance, the names of buttons, default text, labels, and similar types of strings can be found in this file. This file is fully responsible for their content.

The R File

This file is the paste between the activity Java files and the resources like strings.xml. To keep your application run safely, you don't dare to change the file.

The Layout File

This activity_main.xml file is actually a layout file which is available in res/layout directory. It is referenced by your own application when building its user interface. You will change this file very commonly to modify the layout of your application. This file will be shown in the screen. You have various opportunities to change and create a new environment and features of your own through different tools. There are various types of layout format. You can use any of them you wish. You can also change their height, width and other attributes quite easily.

Design is the representation of your app. You can use various attributes and colors to make your design more reliable and efficient. So, people can like it and get easy to use it. There are various kinds of tags such as

<TextView></TextView>,   <Button></Button>,

<RelativeLayout>   </RelativeLayout>,

<LinearLayout></LinearLayout> etc.

Running the Application

Let's aim to run our Hello World android application, we just made. By following this report you can easily create your own AVD where the result will display. To start the app, open your project's activity files (one of your project files) and click Run icon. In another way, you can right click the android project and then run option and then click android application. Eclipse then installs the app on your Android Virtual Manager and initiates it and if everything is ok with your install and application, "Hello world!" will show on your Emulator window.

Conclusion

Though Google doesn't start application development first in android, it creates, builds and takes this platform in a situation where everyone can learn and develop an app and build his/her career so efficiently that he/she doesn't need to look back. This book has described some basic fundamentals of Android Application Development for beginners who want to become an android developer and programmer.

I want to personally thank you for reading my book. I hope you found information in this book useful and I would be very grateful if you could leave your honest review about this book. I certainly want to thank you in advance for doing this.

If you have the time, you can check my other books too.

www.ingramcontent.com/pod-product-compliance
Lightning Source LLC
Chambersburg PA
CBHW070905070326
40690CB00009B/2009